I CAN BE A

WEATHER FORECASTER

℗ CHILDRENS PRESS ®

CHICAGO

Library of Congress Cataloging-in-Publication Data

Martin, Claire.
  I can be a weather forecaster.

  Includes index.
  Summary: Introduces the job of weather forecasting,
describing some of the methods used to learn about
atmospheric conditions.
  1. Weather forecasting—Juvenile literature. 2. Meteorologists—
Vocational guidance—Juvenile literature. [1. Weather
forecasting—Vocational guidance. 2. Meteorology—Vocational
guidance. 3. Vocational guidance.]   I. Title.
QC995.43.M37   1987   551.6'3'023   86-31763
ISBN 0-516-01908-2

Childrens Press, Chicago
Copyright © 1987 by Regensteiner Publishing Enterprises, Inc.
All rights reserved. Published simultaneously in Canada.
Printed in the United States of America.
1 2 3 4 5 6 7 8 9 10 R 96 95 94 93 92 91 90 89 88 87

PICTURE DICTIONARY

weather forecaster

weather report

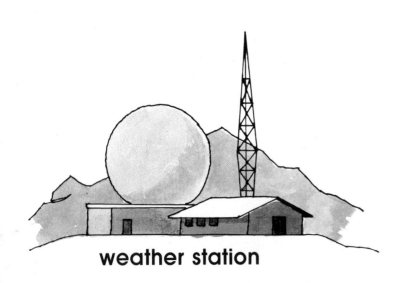

weather station

thermometer

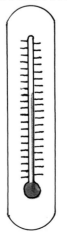

barometer

weather satellite

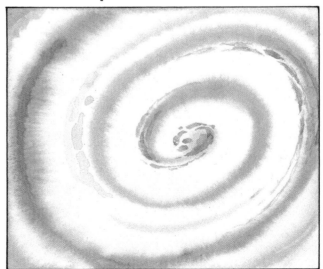

balloon

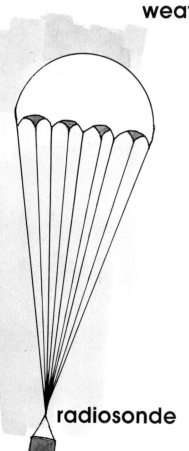

radiosonde

satellite picture

You are a weather forecaster! Everyone is, at one time or another.

Suppose you and some friends are going on a picnic tomorrow. . .if it doesn't rain. You look up

weather forecaster

at the sky. The clouds are light and fluffy, not stormy-looking. The sun is shining, and there is a gentle breeze.

"I don't think it will rain tomorrow!" you tell your friends. You have just given a weather forecast.

Of course, you can't see what's happening many miles away. A faraway storm could be blowing your way.

The air is like the ocean, always moving. In some places it is calm and peaceful, while in others it whirls and twists.

How hot or cold the air is, how heavy or windy or wet it is. . .that is what we call weather.

weather report

To get a better weather forecast, you turn on the radio or television for the weather report.

Some TV weather reporters are meteorologists and some are not.
But all of them try to show us what the weather is doing and why.

The weather reporters
on radio or TV usually
have a good voice and
a personality that people
like. Sometimes they are
funny. Sometimes they
are serious. Their job is
interesting, because
weather is always
changing.

Many weather reporters
are meteorologists, but
not all of them are. The

Meteorologists use many computerized instruments.

reporters you hear and
see on radio and TV
may not be. They may
have gotten their
weather information from
a meteorologist.

Weather maps posted in a large international airport

Meteorologists are
weather scientists. They
study science and
mathematics in college.
They spend years
studying the atmosphere,
the air surrounding the
earth.

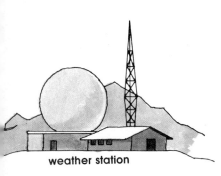

weather station

weather satellite

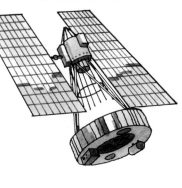

Most meteorologists in the United States work for the National Weather Service. Some companies hire their own meteorologists.

When meteorologists prepare their forecasts, they get lots of help from other meteorologists. Weather stations and weather satellites gather information, which is put together on computers.

Above and top: These National Weather Service centers are receiving weather reports from meteorologists all over the country. Left: This "PROFS" weather system is showing wind speeds picked up by radar.

These are two types of barometers. The one on the
right shows constant changes in the air pressure.

Then the information is
sent swiftly to the
forecasters.

Meteorologists use
weather instruments.
Some of these, like the
thermometer and
barometer, have been
used for hundreds of
years.

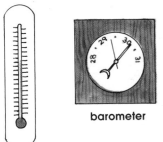

barometer

thermometer

A thermometer in the shade will register a lower temperature than one in the sun.

The thermometer tells how hot or cold the air is. The barometer tells how heavy it is. Did you know that air has weight? It does. And if the air suddenly gets lighter, and the barometer shows the "barometric pressure" falling rapidly, then storms are likely.

We like to know when an icy storm is coming so that we can prepare for it.

Weather forecasts are usually right, but not always. Sometimes it is not possible to tell exactly what the weather will do.

The weather report is important to many

Bad weather can ruin our fun—
or sometimes even our homes.

people. Almost everyone
wants to know if it will
rain. Farmers need to
know about rain, heat, or
cold. Sailors and pilots
must know if storms are
on the way.

Damage left by a tornado (left), which travels as a funnel-shaped cloud (right)

Even before airplanes were invented, scientists wanted to know what was happening in the atmosphere. Once, two men took thermometers, barometers, and other weather instruments high into the air. They took

A spectacular lightning storm

them up in a big hot-air
balloon! They learned a
lot about what happens
in the atmosphere. But
they also learned that it
was dangerous to go so
high in a balloon. They
almost died from cold
and lack of air.

balloon

radiosonde

Weather forecasters still use balloons to learn what goes on in the atmosphere. They send packages of instruments called "radiosondes" into the air, attached to balloons. But people don't have to go up with them! The radiosondes send radio signals back to weather forecasters on the ground. The signals give information about

Left: A meteorologist receiving data sent from a radiosonde
Right: A meteorologist launching a radiosonde balloon

the temperature, humidity,
and barometric pressure.

Thousands of radiosondes
are released every year.
They return to the earth
on small parachutes. Some
are lost. But others land

This weather satellite takes photographs that help meteorologists understand weather conditions.

in people's yards, fields, gardens, or woods. If you find a radiosonde, send it to the address on the package. The instruments may be repaired and used again.

Today, using weather satellites, forecasters can see storms developing.

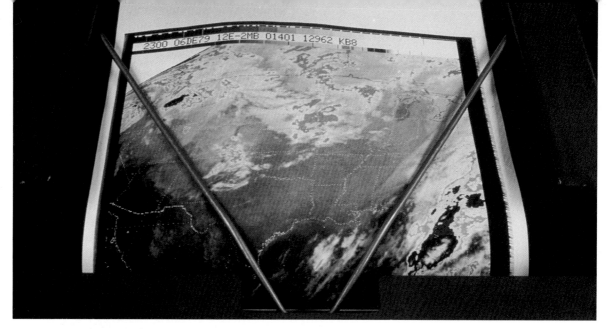

A satellite photo just coming in to a weather station

Weather satellites orbit
above the atmosphere
and look down on our
ocean of air. They send
pictures back to the
earth. You often see
satellite pictures on the
TV weather report.

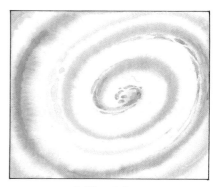

satellite picture

Satellite pictures are
very important in spotting

Left: This hurricane reporting center gives up-to-the-minute reports.
Right: The National Hurricane Center in Miami, Florida

hurricanes. Hurricane forecasters study the satellite pictures. When they see a hurricane forming, they warn people who might be in its path. Hurricanes are the most dangerous of storms.

Above: Hurricane-damaged homes near the Mississippi coast
Below: A hurricane swirling over the Atlantic Ocean

This special airplane spots hurricanes and reports them.

Some meteorologists fly in special planes right into the center of a hurricane. Violent winds bounce the planes around. Radar and computers on board give the forecasters valuable information about the strength and direction of the storm.

The National Weather

This weather observatory atop Mount Washington in New Hampshire measured a wind chill index of minus 112 degrees Fahrenheit (minus 80 degrees Celsius).

Service uses special radio broadcasts to warn people about other emergencies, like tornadoes or flash floods.

During weather emergencies forecasters may have to go for days with little sleep. Some weather stations are in lonely areas, and the

forecaster may work alone. In the future, there could even be forecasters in space stations orbiting above the earth.

Do you like to study about the world you live in? Do you like science and arithmetic? Maybe you would like an interesting and exciting job like weather forecasting.

WORDS YOU SHOULD KNOW

atmosphere (AT • mus • fear)—the air surrounding the earth

barometer (buh • RAH • muh • ter)—an instrument used to measure the weight (pressure) of the air. The barometer contains a thin tube of mercury. As air pressure rises and falls, the level of the mercury in the tube rises and falls.

barometric pressure (BARE • uh • met • rick PRESH • er)—the weight (pressure) of the air, as shown by the level of mercury in a barometer

flash flood (FLASH FLUD)—a very large flood that starts and ends quickly, caused by heavy rainfall

forecaster (FORE • cast • er)—a person who calculates or predicts future events

humidity (hyoo • MID • ih • tee)—the wetness of the air

hurricane (HER • ih • cane)—a tropical storm with high-speed winds

meteorologist (mee • tee • er • OL • uh • jist)—a scientist who studies the weather

personality (purse • un • AL • ih • tee)—the style in which a person speaks, acts, and reacts

radiosonde (RAY • dee • oh • sahnd)—a small radio transmitter that is carried into the atmosphere by a balloon and sends back weather information

satellite (SAT • uh • lite)—an object that circles around a planet or other heavenly body

thermometer (thur • MAH • muh • ter)—an instrument that measures the temperature

tornado (tor • NAY • doe)—a windstorm that travels over land as a funnel-shaped cloud

weather (WEH • ther)—what is happening in the atmosphere, such as heat or cold, wetness or dryness, calm or storm

INDEX

PHOTO CREDITS

ABOUT THE AUTHOR

Claire Martin graduated from Mount St. Mary's College in Los Angeles and received a Masters Degree in Library Science from Immaculate Heart College, also in Los Angeles. Her interest in writing for children began when she was raising her own seven, and continued as she worked as a children's librarian for the City of Los Angeles. She and her husband now live in New Jersey, where she is a reference librarian at Morris County College and an author of children's books.